Comb Management

by Wally Shaw

Northern Bee Books

Comb Management
© Wally Shaw

All rights reserved. No part of this publication may be reproduced, stored in a retrieval system, transmitted in any form or by any means electronic, mechanical, including photocopying, recording or otherwise without prior consent of the copyright holders.

ISBN 978-1-912271-35-1

Published by Northern Bee Books, 2018
Scout Bottom Farm
Mytholmroyd
Hebden Bridge HX7 5JS (UK)

With the agreement of Cymdeithas Gwenynwr Cymru
The Welsh Beekeepers Association.Bridge HX7 5JS (UK)

Design and artwork by DM Design and Print

Printed by Lightning Source UK
Printed by Lightning Source UK

Comb Management

by Wally Shaw

Contents

Introduction	1
Measurements	2
Wild Nests	2
How does a Movable Frame Hive Differ from a Wild Nest	4
Which of these Differences are Important for the Bees?	5
Methods of Frame Spacing	8
Things NOT to do with Frame Spacing	10
Frames	11
Frame and Spacing Summary	12
Comb Drawing	13
Drawing of Brood Comb	14
Drawing of Honey Comb	16
Drawing Comb with a Swarm	18
Appendices	19
Appendix I - The Demaree Method	21
Appendix 2 - The Bailey Comb Change	24
Appendix 3 - The Shook Swarm	29

List of Figures

Figure 7: Inter-comb spacing using frame spacing of 35mm	4
Figure 2.· Alternative methods of frame spacing	7
Figure 3: Frame types	11
Figure 4.· Comb drawing early in the season	15
Figure 5.· Comb drawing later in the season	16
Figure 6: Demaree method of comb replacement	22
Figure 7. The Bailey comb change	25
Figure 8.· DIY Bailey Board	25
Figure 9.· A shook swarm	29

Comb Management

Introduction

Honey bees can successfully live in all sorts of different nest sites – a hole in a tree, a chimney pot or a bee-hive – but in all cases this is just a cavity in which to make a set of combs. It is in and on these combs that all the within the colony functions occur. Because it is dark in the hive, communication is through pheromones or vibration and combs provide the ideal carrier for this information. For example, bees can always locate the queen by following the trail of her footprint pheromone on the combs. The main outside the hive activities are foraging, swarming and queen mating. As beekeepers, interested in the production of honey, we tend to concentrate on the foraging activities of our bees and it is easy to overlook the fact that over 95% of a typical worker bee's life is spent within the confines of the colony engaged in some activity in or on the combs. In a sense, the combs are an extension of the bees that made them and it is bees and combs together that constitute the colony.

Up until about 1850, bee colonies, whether wild of under human stewardship (it hardly qualified as management), built themselves a set of combs entirely according to their own design in whatever cavity they could find or was provided by the beekeeper. No restriction was placed on the way the colony used these combs to engage in their main activities of brood rearing and food storage. With the introduction of the moveable frame hive, followed quickly by the invention of wax foundation and the queen excluder, everything changed. Beekeepers were now able to induce the bees to make their combs where they (the beekeepers) wanted them, ie in wooden frames. The beekeeper could now even influence the size of cells they built by the dimensions of the hexagon embossed on the sheet of wax. It also became possible to separate the use of combs for brood rearing and honey storage using a queen excluder. Some of the changes that modern beekeeping has imposed on colonies have potential effects on the health and welfare of the bees and others do not.

Measurements

We are now going to discuss the dimensions of combs and their spacing and this is most easily expressed in metric units (millimetres or mm) rather than Imperial measure (inches and fractions of an inch). A difference of one millimetre (1mm) may seem insignificantly small to you but it is a lot to a bee – equivalent to over 1 foot or 300-400mm for us humans! So, if you still think in Imperial, here are a few conversions to help you visualise things:-

¼ inch = 6.5mm   ⅜ inch = 9mm   ½ inch = 12.5mm   1 inch = 25mm   1¼ inch = 31mm   1⅜ inch = 35mm   1⁷⁄₁₆ inch = 36.5mm   1½ inch = 38mm   1⅞ inch = 47.5mm and 2 inch = 51mm

Bee space is ¼ inch - ⅜ inch or 6.5-9mm. Bee space at the bottom of the combs is different at 15-16mm.

Wild Nests

When a swarm of bees takes up occupancy of a nest cavity their first task is to build a set of combs. No matter where they build, the spacing of the combs will always have a very regular spacing of 30-32mm (between centres). Only combs at the outside of the nest may have slightly wider spacing but these are used almost exclusively for honey storage or drone brood. Depending on the shape of the cavity, the combs they build are rarely flat and are often curved in graceful arcs. They may also be joined in places and braced to the cavity wall. Bracing increases structural integrity so that the combs can bear the weight of a full load of honey - even in hot weather.

In terms of usage, the combs in a natural nest are multi-purpose and brood is raised almost anywhere with the possible exception of the outside combs. The brood nest tends to start high in the spring and work down during the season. When the nest starts to contract in the autumn cells from which bees have recently emerged are back-filled with the over-wintering stores. In a wild nest the queen has access to more comb than when restricted by a queen excluder and moving from one comb to another is part of their hygienic behaviour. The practice of 'nadiring' in Warré hives, where new boxes with top-bars on which the bees build new combs are added at the bottom and brood rearing is not restricted by a queen excluder, is closer to a natural nest. In conventional hives an acceptable substitute can be provided by regular comb replacement. The ultimate method of doing this is the Bailey Comb Change or the Shook Swarm (see **Appendix 2 and 3**).

It is important to understand why natural comb spacing is 30-32mm. Measuring cells that have been prepared for the queen to lay shows that they are all 11-12mm deep. More detailed measurements reveal that it is not uniform depth that the bees are actually targeting but rather -a width/depth ratio of about 1:2 – so smaller (diameter) cells are slightly shallower and larger cells are deeper. Why this particular ratio? The only time a cell is fully occupied is in the later stages of brood development and these are the proportions that exactly fit the occupant – the pupa or pre-emergent bee. Smaller (diameter) cells produce smaller bees; hence it is the width/depth ratio not the absolute depth that is important. Taking these calculations a stage further, and assuming an average cell depth of 11.5mm, the comb width, where both sides are prepared for brood raising, is 23mm (2 x 11.5). Subtract this from the natural comb spacing of 32mm (32 – 23 = 9) and this leaves a space of about 9mm between opposing comb faces (see **Figure 1**). It is no coincidence that this is exactly the right space to allow a layer of worker bees on opposing comb faces with their wings just brushing. This spacing allows the nurse bees to move about freely to gain access to the cells for tending the brood. The provision of a 9mm space is also important for thermoregulation which can be most efficiently achieved by just two layers of bees between the combs. In colder conditions more bees are recruited to occupy the comb inter-space where they both generate heat and slow the convective flow of air. If the bees are forced by the beekeeper to make combs at wider spacing it requires more bees – and more than two layers - to be packed in under cold conditions to keep the brood warm. The bees are fully aware of the need to keep brood warm under all circumstances and wider spaces mean that less brood will be produced.

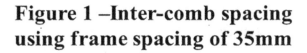

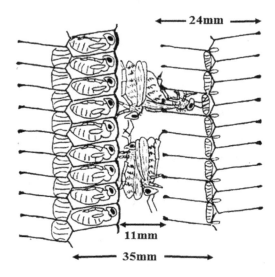

Figure 1 – Inter-comb spacing using frame spacing of 35mm

When brood combs (or parts of them) are used for honey storage at the end of the summer (in preparation for the winter) the cells are extended to leave a single bee space between them (5-6mm), just like in the honey supers. But, before these cells can be re-used for brood next spring, they have to be trimmed to the correct depth. This activity accounts for the drifts of wax fragments that are found under open-mesh floors at this time of year.

How does a Movable Frame Hive Differ from a Wild Nest

Until the movable frame hive came into widespread use, and colonies were kept in skeps (or various other containers), the bees were able to live in almost exactly the same way as in the wild. They built and used combs in an entirely natural way, the only difference being that a set of combs rarely had more than a 1-2 year life before the beekeeper killed the colony in order to harvest the honey. At harvest the combs were cut out of the skep and, when it was re-occupied by a swarm next year, they had to re-build from scratch. This meant that normally no combs were more than 2 years old – the healthy option! Even before Varroa came on the scene, wild nests rarely remained in continuous occupation for more than 5-6 years. As soon as the nest cavity was abandoned, wax-moth would move in and destroy all the old combs so that when it was re-occupied by a swarm the bees had to build new combs from scratch.

The introduction of the movable frame hive imposed some very big changes in the ways the bees live as compared with their natural way of doing things. The most significant changes can be summarised as follows:-

a) The bees are forced to build their combs based on sheets of embossed wax foundation (usually with embedded reinforcing wires) firmly fixed in wooden frames.

b) The use of a queen excluder inhibits natural use of the combs for brood rearing and the queen is often forced to lay in the same old combs year in year out.

c) Because the hive is now modular, the beekeeper can increase the space available for both brood and honey by strategically adding new boxes of comb. This delays or prevents swarming and creates much larger colonies than would normally exist in nature.

Which of these Differences are Important for the Bees?

They are all important for the beekeeper (which is why we do them) but how does it affect the colony? The most important difference is the extended life of framed, reinforced combs which are so robust that they can be re-used over many years. The use of plastic foundation potentially makes this problem even worse. When the bees are forced to use the same old brood frames year after year this has serious implications for disease control. **For this reason, regular replacement of brood combs should be standard practice.** In areas where brood disease is not a problem, comb renewal on a 3-year cycle is adequate. If disease is a problem locally, then more frequent changes are advisable. However, when the queen has access to new combs she lays more, resulting in a larger colony, so the cost of foundation is usually recovered by a larger honey crop.

The use of a queen excluder compounds the problem of old combs by preventing the queen laying in new combs higher up in the hive. The extended life of honey combs – those that have never been used below the queen excluder – is not usually a problem and they only need replacing when damaged. An outbreak of disease may sometimes require them to be sterilised.

Our old friend the wax-moth is definitely on our side here, or at least this is how they should be regarded. When combs that once held brood, and have the remains of pupal skins in their cells, are stored outside the hive they become vulnerable to wax moth which quickly renders them useless for further service – which is good news in some cases. Super combs that have never held brood may have wax-moth eggs laid on them but, because there are no pupal skins to provide a source of protein, the young larvae will not survive long enough to do any significant damage. However, super combs that contain stored pollen are more vulnerable.

Apart from longevity, the fact that the comb is constructed within a wooden frame with reinforcing wires has few disadvantages for the bees (but huge advantages for the beekeeper). You only have to try and examine the free-hanging combs in a top-bar hive to realise just how fragile they are. The wooden frame makes inspection of the hive much easier and is absolutely essential for the extraction of honey without destroying the combs. The only disadvantage for the bees is that framed combs do not transmit vibrations as well as free-hanging comb. Although we humans perceive the bee foraging dances visually, for the bees it is quite different. In the dark of the hive, the information contained in the dance is communicated to the onlookers in the form of vibrations transmitted by the comb. To overcome this deficiency the bees will often remove a strip of wax along the bottom-bars and part way up the side-bars so that these combs are better sounding boards. Not understanding the reason, beekeepers often view this behaviour in an unfavourable light. Such `dance floor` combs are usually found in the middle of the bottom brood box and should be left in this position if possible.

Neither the much larger colony that is produced in a modular hive (as compared with a natural nest) nor the fact that the combs are built in a different order, seems to present any significant problems. Each box of comb is built from the top down and this seems to be acceptable to the bees – except that they will always try to create comb continuity between boxes if the bees-space between them is not right (and sometimes even when it is). In nature bees build comb in all sorts of different cavities and have to be very flexible in how they build their nest to accommodate this variability.

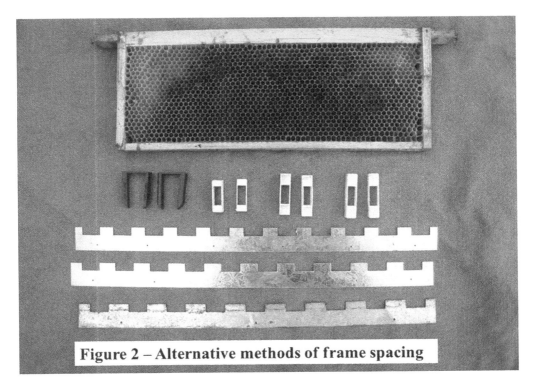

Figure 2 – Alternative methods of frame spacing

Methods of Frame Spacing

In Britain we are blessed (or blighted) by having no less than three different methods of frame spacing in common use (see **Figure 2**):-

1) **Hoffmann self-spacing frames** – giving 12 frames/box (National hive). The Manley is also a self-spacing frame ($1^5/_8$ inch or 41.5mm) but is only for use in honey supers.

2) **End spacers** – these used to be made of folded metal but are now mostly plastic. Plastic end spacer come in no less than 3 sizes; narrow spacers ($1^7/_{16}$ inch or 36.5mm), wide spacers ($1^7/_8$ inch or 47.5mm) and extra wide spacers (2" or 51mm). These different sizes enable a variety of frame spacing from 8-11 frames/box to be attained. Narrow spacers are used end-to-end only but the two wider varieties can be used end-to-end or (staggered) overlapping if required.

3) **Castellations** – are available in 3 versions of 9, 10 and 11 frames/box (12 frame castellations are not commonly used but are said to be available to special order). Castellations for WBC hives are available in 8, 9 and 10 frame spacing.

This choice of methods is somewhat confusing, especially for the beginner beekeeper, so here are a few guidelines:-

▸ Brood frames are best spaced at 35mm (12 frames/box) as this is closest approximation to the natural comb spacing of 32mm.

▸ Brood frames should not be spaced wider than 37mm (11 frames/box) or the space between comb faces becomes too wide when they are being used for brood.

▸ Frames containing foundation (for drawing) should also be spaced no wider than 37mm (11 frames/box) – wider spacing usually results in irregular drawing of combs.

- Drawn frames for honey storage can be spaced at anything from 11 frames/box (37mm) to 8 frames/box (51mm). The wider the spacing the more likely the bees are to build brace-comb and 8 frames/box usually proves to be a bit too extreme.

Hoffmann Frames

These self-spacing frames have a fixed spacing of 35mm (34.5mm if using plastic converters) which is ideal for brood rearing. In many countries this is the only method of spacing used throughout the hive. Interestingly, other countries have Hoffmann frames that provide narrower spacing (down to 33mm) than those available in Britain. Hoffmann frames are not ideal for honey production because they contain less honey/frame and may have areas of capping that are below the level of the surrounding frame – impossible to uncap with a knife. Hoffmann frames are also not ideal when using a tangential extractor because as they do no lie flush (comfortably) on the wire screens.

The big advantage of Hoffmann frames is that once the first frame (or dummy board) has been removed from the box, they can be moved laterally (slid along the frame runners) in groups in order to access particular frames for examination without disturbing the other frames. They can also be slid back into place in a similar manner but must be pushed together tightly in the hive to avoid the build-up of propolis on the contact faces. The ideal set-up for a brood box is probably 11 Hoffmann frames and a dummy board. The latter can be used to lever the frames tight after they have been disturbed during inspection.

End Spacers

This is the most flexible method of frames spacing because a box can be re-configured by a change of end spacers. This is useful to beginners who may have a limited amount of equipment. A range of frame spacing can be achieved from 8 frame/box (51mm) to 11 frames/box (37mm). The wider the frame spacing the more it invites the bees to build brace comb. As it is not possible to get narrower than 37mm using end spacers, they are less than ideal for brood frames. End spacers also have the infuriating habit of either falling off at the wrong moment (usually into the hive or long grass) or being stuck solid with propolis when you want to change spacing. Another factor to bear in mind is that, if (or when) we get small hive beetle in Britain, end spacers may become a liability because they will provide nice little hidey-holes where beetles can lurk safe from molestation by the bees.

Castellations

Using castellations means having boxes dedicated to a given frame spacing, so it is not as flexible as end spacers. They are not ideal for brood because during inspection frames initially have to be lifted vertically (10-12mm) before they can be moved laterally. The same manoeuvre in reverse is required (individually) to put the frames back. This tends to roll bees between the comb faces - and a rolled bee is not a happy bee! Castellations are ideal for honey supers because the frames are held very firmly and do not rub together and leak honey whilst in transit. This method of frame spacing minimises the use of propolis.

Things NOT to do with Frame Spacing

- Do not have a mix of Hoffmann and end spacer frames in the same box; they do not mix well. The variable spacing tends to produce irregular combs that can not be easily repositioned within the box during hive manipulations – they may only fit in one place! Buy yourself some plastic converters to transform ordinary frames into Hoffmann frames and have done with it!

- Do not leave gaps in boxes of comb or you will get brace comb or combs bulging on one side that will only fit in one place in the box. Get the spacing right or use a dummy board!

Frames

For the Modified National and WBC hives, Standard British (BS) frames are sold in a range of types using a code name. When purchasing equipment it helps to understand what these codes mean. The first element of the code is the depth of the frame (to fit shallow or deep boxes) and **S** = shallow and **D** = deep. This followed by the letter **N** which stands for National and this is followed by the numbers 1-5 (except that number 3, with a slot in the top-bar, is no longer produced)). See **Figure 3**.

SN1 and **DN1** frames have $7/8$ inch wide top-bars and (plain) $7/8$ inch side-bars.

SN2 and **DN2** frames have wider $1 1/16$ inch top-bars but the same $7/8$ inch plain side-bars.

SN4 and **DN4** frames have $7/8$ inch top-bars but Hoffmann 35mm self-spacing side-bars.

SN5 and **DN5** frames have the wider $1 1/16$ inch top-bars and Hoffmann 35mm side-bars.

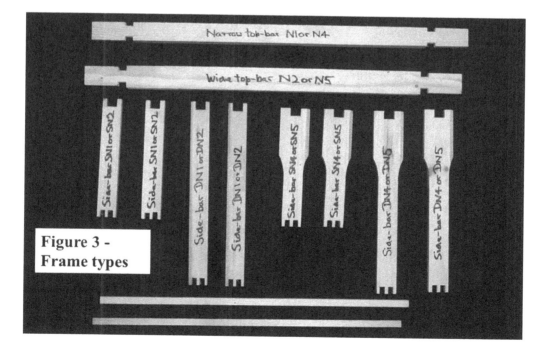

Figure 3 - Frame types

There is an identical range of BS frames for the Smith hive, except that the Smith top-bar has shortened lugs. Commercial, 12 x 14 (an extra-deep Modified National box), Langstroth and Modified Dadant hives all have their own specialised frames. A wider (1½ inch or 38mm) Hoffmann side-bar is also offered for all these frames but is rarely seen.

What's the difference? For Hoffmann frames the difference is self-evident – they are self-spacing. Hoffmann frames can also be made by converting non-Hoffmann frames using a nailed-on plastic converter. The idea behind the wider $1^{1}/_{16}$ inch top-bar is that it provides the correct bee-space between top-bars and thus reduces the amount of brace comb built in this position. It does work and, for brood frames, the extra cost is probably worthwhile – but probably not for super frames. However, if you already have a large stock of frames with narrow top-bars do not despair (keep your wallet closed), they are still perfectly serviceable – you just get a bit more brace comb.

Frame and Spacing Summary

Much has been said above about the importance of correct frames spacing but, if you are going to use a moveable frame hive (there's a clue in the name) in which the frames actually move (freely), the hive boxes must fit together and leave correct bee-space (¼ inch - $^{3}/_{8}$ inch or 6.5-9mm) between them. The space in question is that between the bottom-bars of the frames in the box on top and the top-bars of the frames in the box below. Whether you use a hive with bottom bee-space (standard for the National hive) or top bee-space (Smith, Langstroth and Dadant hives), it is vital to check this space is correct for all boxes. If in doubt you can check a bottom bee-space box by placing it on a flat surface and hanging a frame in it. You should just be able to insert your little finger in the space beneath the bottom-bar.

Standardisation is important in beekeeping and arguably the best choice is Hoffmann frames (for brood) below the queen excluder and castellations above (the honey supers) - with a few supers using end spacers to provide flexibility. At least some of the castellated supers must be of the 11 frame variety in order to get foundation drawn properly. Wider spacing in honey supers gives more weight of honey/box and fewer frames to extract. On the other-hand, narrow spacing (with more frames) gives more cells for the bees to work on when depositing and drying nectar and also gives more surface area for passive drying prior to capping. **It is for brood frames that correct spacing really matters.** Spacing of super frames is only critical for comb drawing.

Comb Drawing

Beekeepers often put frames of foundation in a hive and are disappointed when they do not get drawn. A colony will only draw combs when it actually **needs to** – when there is a shortage of comb for what they want to do at that particular moment. There is no anticipation of future needs and the `here and now` requirement for comb is the only consideration. Comb building is a heavy drain on colony resources and it is estimated that it takes 8lbs (3.6kg) of honey to produce 1lb (0.45kg) of wax. Additional costs are incurred through the diversion of bees from other duties to become comb builders and the energy needed to produce the local increase in temperature (about 42°C) that is required for the bees to work the wax properly. A colony is also more willing to engage in comb building in late spring and early summer – when the queen is laying hard and bee numbers are rapidly increasing – than it is at any other time of the year. The two main reasons why the colony commences comb building are:-

1)   It has run short of space for the brood nest to expand.

2)   It has run out of space to store honey.

In a natural bees nest the distinction between space for brood and honey does not exist and it is only in a movable frame hive with a queen excluder in place that the two reasons need to be considered separately.

Assuming a colony needs to build comb it also has to have the resources to do it. The most important resource is **a current flow of nectar**. Stored food is not normally used for comb building. This is an important survival strategy and conserving existing stores always has priority over comb building. It does not have to be a natural nectar flow and the provision by the beekeeper of sugar syrup is just as effective. The other requirement is plenty of bees to become wax makers. However, if there are not plenty of bees there will be no need for comb building which is why it is a waste of time putting frames of foundation in a weak hive and expecting them to be drawn.

Drawing of Brood Comb

Brood combs are drawn when the brood nest is expanding and more space is required for the queen to lay. When the brood nest encounters a frame of foundation the bees will draw it. However, if there is an unused drawn comb on the other side of the box, the brood nest will usually expand in that direction first. Many beekeepers put frames of foundation on the outside, next to the hive wall, expecting them to get drawn. In this position they will only be drawn as a last resort and even then often only on the inner face. Bees much prefer to build a vertical column of combs than to expand laterally. Changing the position of frames, ie. moving drawn combs to the outside and interleaving frames of foundation with drawn combs, can be used to force the bees to draw combs when and where the beekeeper wants them but it must be done sensibly taking into account the size of the colony and the time of the year.

Early in the season - when the brood area is not yet full of bees and the weather is still cool, the correct place for frames of foundation is next to the outermost frame of brood. Even though there are drawn frames outside them, in this position they will have to be drawn to retain brood nest continuity (see **Figure 4**).

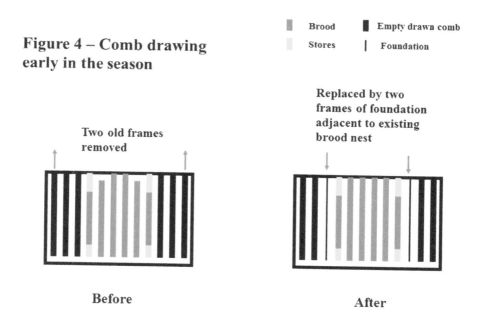

Figure 4 – Comb drawing early in the season

Before

After

Later in the season – when the brood area is crammed with bees, frames of foundation can be placed within the brood nest itself (see **Figure 5**). In this position they will be drawn and probably laid in 4-5 days, or less. It is often claimed that interrupting the brood nest with frames of foundation leads to production of queen cells on the side where the queen is (temporarily) absent but this does not seem to happen in practice. Providing the colony is large enough and the queen is laying hard, a frame of foundation will not act as a barrier and she will happily by-pass it to lay on the other side.

Comb Management

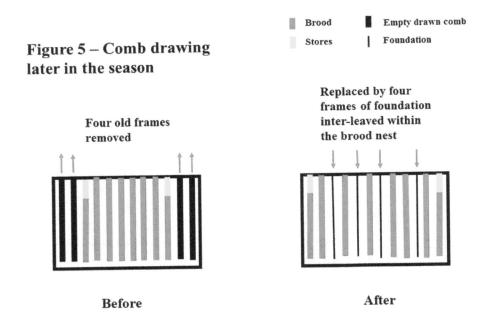

Figure 5 – Comb drawing later in the season

A Word of Warning – if you have put foundation in the brood area and when you come back a week later the first thing you see (without even removing any frames) is that it has not been drawn you will almost certainly find that queen cells have been started. As far as I am aware, this has not been caused by the foundation but simply that the colony was pre-destined (already triggered) to swarm. What's needed now is to artificially swarm the colony.

Drawing of Honey Comb

The building of combs to store honey is less critical for both the bees and the beekeeper. Badly drawn combs with depressions and bulges will hold just as much honey as a pristine set of regular, parallel combs. The bees will be quite happy with their handiwork and it is only the beekeeper that will suffer when extraction time arrives. So to avoid a lot of mess and poorly balanced loads in the extractor it is better to pay some attention to getting nice evenly drawn honey combs.

Beekeepers often put supers of foundation on a hive and are then disappointed when it is not immediately drawn. However, the initiation of comb building for food storage is well understood. Incoming foraging bees do not store nectar themselves; they transfer it to receiver/storage bees somewhere near the hive entrance. When these bees are fully loaded they move off to find somewhere to store the nectar. If after a reasonable time ('reasonable' is not specified in the literature but it is probably minutes rather hours) the storage bee has not found a vacant cell in which to deposit the nectar - one that is not already full of nectar or is not currently being used by another bee – it ingests the nectar. In this process, the nectar passes from the bee's honey crop into its digestive system and this large intake of sugars converts it into a wax maker. Switching usually commences when 60-70% of the available storage space is already filled with nectar. Having converted to a wax making and comb building duties, this bee will repeatedly visit the hive entrance to re-fuel from incoming foragers and make more wax. Each cycle of wax making (8 scales) takes about 12 hours. Only when the nectar flow ceases will this behaviour be switched off. The need for new comb will have ceased with the nectar flow so this provides a self-regulating mechanism for the control of comb building – so they do not build more comb than they need.

There is a range of ideas about how best to present combs to a colony to ensure good drawing but it is not clear that this is really important – it is the colony's need to make new comb and a nectar flow to fuel it that really matters. Generally comb building is easiest for the bees in the warmest place in the hive and that is immediately over the queen excluder. But, if the weather is warm and there is a good nectar flow, they will happily draw comb at the top of the hive. Placing a frame containing nectar in the middle of a box of foundation ('seeding') is said to encourage the bees to use the box and draw the foundation earlier than they would otherwise.

Drawing Comb with a Swarm

This is the ideal method of getting combs drawn quickly and well. A natural swam is already well prepared - with many of the bees carrying mature wax scales - to build comb. Even though there may be a nectar flow when the swarm is hived, it is still a good idea to supplement this with a generous feed of sugar syrup – 4L of syrup in a contact feeder is no too much for a good sized swarm. It is often recommended that you leave a swarm 24-48 hours to settle in before feeding. Forcing them to use the honey stored in their crops first may help purge them of disease but I am not convinced this is really necessary.

A swarm should be initially installed in a minimum volume hive and given only those frames that are going to make its brood nest – **NO** supers. The more bees you can cram into a limited space the quicker and better the combs will be drawn. A good sized prime swarm will be quite happy in a single deep box until it has drawn its first set of combs. Do not forget that this colony will only get smaller for at least 3 weeks, until the first brood laid by the queen emerges. If it is a cast swarm, that has to get its virgin queen mated first, it may be 5-6 weeks before the colony gets any new recruits. If you use shallow boxes for brood rearing, then small swarms and cast swarms can be conveniently hived in one of these. It is not a bad idea to give a newly hived swarm one clean drawn frame in the middle of a box of foundation so that the queen can start to lay immediately.

When a swarm occupies a box containing foundation they will start to draw as many frames of foundation as there are bees to cover them. A really large swarm will simultaneously draw all the frames in a deep box in as little as 48 hours. Smaller swarms will only attempt to draw some of the frames (4-8, depending on the number of bees) and will leave the rest untouched. When they have drawn this initial set of combs they will start to use them for brood and food storage and will not attempt to draw any further foundation until they need to – which is when they have fully utilised the initial set of drawn combs. This is not what the beekeeper wants, particularly if syrup is being fed to help with comb drawing. Instead of being used for drawing, this syrup will be stored in the initial set of combs and could later find its way into the honey supers. To avoid this problem, the frames should be re-arranged at regular intervals until all frames are drawn. Frames that have been drawn but have not yet been laid in by the queen (you must check there are no eggs) can be moved out and replaced by frames of foundation. This way all the frames will be drawn in the shortest possible time and the syrup feed can be withdrawn to prevent its use for other purposes. As soon as the initial box of combs is fully drawn and is being well utilised by the bees, other brood boxes or honey supers can be added – but this is often not until the colony size is starting to increase again (3 weeks + later).

Appendices

Appendices 1-3 deal with increasingly more radical methods of renewing the brood combs in a hive. These are of particular use when over a number of years good frame management has lapsed and most or all of the brood combs are no longer fit for purpose. Methods 2 and 3 are also the types of comb management that should be used if brood disease is thought to be a risk.

1. **The Demaree Method** – this enables frames containing brood to be moved to the top of the hive (where the brood can emerge and not be lost to the colony) and be replaced by frames of foundation (see **Figure 6**). This management also doubles-up as pre-emptive swarm control.

2. **The Bailey Comb Change** – this enables a complete new box of frames to be drawn so that most or all of the existing brood combs can be removed (see **Figure 7**). As with the Demaree method, all the existing brood is permitted to emerge and join the colony.

3. **Shook Swarming** – this is the method used by the Bee Inspection Service (APHA) to deal with the less serious cases of European Foul Brood (EFB). In this method all the existing brood is destroyed (lost

Comb Management

to the colony). See **Figure 8**. If EFB has been found in an apiary it is recommended that all hives (even those showing no sign of infection) should be shook swarmed as an effective preventive measure.

NB - All the above methods require the beekeeper to find the queen. Some people find this difficult (sometimes it IS difficult). Advice on this can be found in the WBKA Booklet, *'Simple Methods of Making Increase'*, pages 21-22.

Comb Mangement – Appendix 1

The Demaree Method

Part of good comb management is looking ahead to next year's season and in autumn moving frames that are need replacing to the outside of the box. In this position they are unlikely to contain anything of importance (and particularly brood) during the early part of the season in the following year. This means they can be removed with impunity and replaced by frames of foundation. This should not be a straight swap because the frame of foundation needs to put where it will be drawn almost immediately. This management is usually being done early in the season so the correct place will be adjacent to the current brood nest (see **Figure 4**).

However, with the best will in the world, it is often impossible to deal with all the frames that need replacement in this manner. In some cases there may be quite a lot of them and this is where the Demaree method is useful because frames containing brood can be transferred to a new box on top of the hive (on top of the supers) and replaced with foundation. This means that none of the bees that will emerge from those brood combs will be lost to the colony (you are not weakening it in any way).

The Demaree method is also a very effective means of pre-emptive swarm control. This is because the brood frames at the top of the hive attract nurse bees up from the bottom of the hive, thus reducing congestion in the main brood area. Having to draw foundation is another disincentive to swarming and new combs provide space for the queen to lay – the combined effect of all three is a strong disincentive to swarming.

The method is best applied to a fairly large colony (see example in **Figure 6**), in other words there are a lot of bees in the hive. This means that the best

Comb Management

time is likely to be mid-May to mid-June, when the colony is at or near its peak. Later in the season, as the brood nest starts to contract, the colony will be less willing to draw foundation. As it is not possible to feed a colony in this hive configuration (with supers on it), the comb drawing must rely on there being a nectar flow at the time. It does not have to be a large flow but there has to be some incoming nectar.

The example shows a hive on a brood and half and with two part filled supers. Four frames containing brood have been removed from the deep brood box at the bottom of the hive and placed in a new deep box on top (above the supers) where they are flanked by dummy boards. In the bottom box, the removed frames have been replaced by frames of foundation which are interleaved in the remaining brood nest. Three or four frames is usually all that is necessary but in needy cases the process can be repeated after an interval of 7-10 days – by which time the foundation at the bottom of the hive should have been drawn and in use. So you can end up with 7-8 frames in the box at the top of the hive.

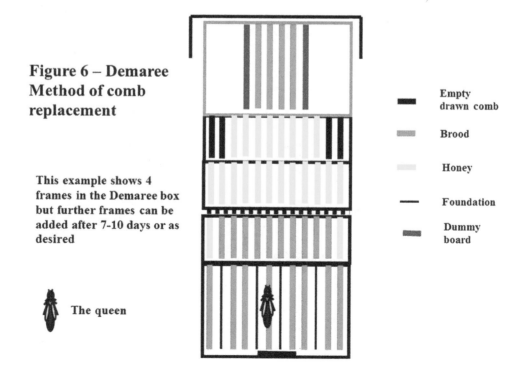

Figure 6 – Demaree Method of comb replacement

This example shows 4 frames in the Demaree box but further frames can be added after 7-10 days or as desired

The queen

— Empty drawn comb

▬ Brood

▬ Honey

— Foundation

▬ Dummy board

The following is a step-by-step guide to the manipulation required to do a Bailey comb change (see Figure 8):-

1. Open the existing deep brood and inspect the frames. Any that contain no brood should be removed for destruction or re-cycling. At the same time look for the queen and when found place her in a safe place – a queen marking cage (the type with a foam plunger) comes in useful here.

2. When all the empty frames have been removed, move the remainder (those containing brood) into the middle of the box and flank them with a dummy board on either side.

3. Install the queen excluder/Bailey board on top of the re-arranged old brood box.

4. Place the new deep box of foundation on top of this. It is a great advantage if this contains one good quality, empty drawn frame on which the queen can start to lay immediately.

5. If the hive also had a shallow brood box (a half-brood) this can be placed on top of the box of foundation. Because this box contains some brood and nurse bees it may be a gentler transition for the queen. However, the beekeeper needs to be aware that some of the syrup provided for comb drawing may be stored in this box.

6. Replace the queen in the upper box (don't forget to do this!). She may seem a bit lonely at first but bees will soon move up to join her.

7. Replace the cover board and install a contact feeder (preferably a 4L one) containing a strong sugar syrup (2lbs to 1 pint or 1kg to 625ml water). Replace the roof. The type of feeder used needs to provide a fairly low flow so that the syrup is used for drawing and is not stored in large quantities.

8. In 5-7 days check on progress. Is the queen laying well? Has she started to lay on newly drawn frames? How many frames have been drawn? Does the feeder need a re-fill?

9. About 2 weeks later most or all of the foundation should have been drawn. The bees will often fail to draw frames near the edge of box and these can now be moved into positions adjacent to the current brood nest where the bees will be forced to draw them. You must ensure the integrity of the brood nest, so check that any frames you move out do not contain young larvae or eggs!

10. After 24 days (allowing for the drones to emerge) the bottom box containing the old combs will be brood free and can be removed altogether.

11. The box of new combs, which by now should contain a substantial amount of brood, can now be lowered onto the floor. The feeder can be removed and normal seasonal management (adding supers as required) can now be resumed.

Comb Mangement – Appendix 3

The Shook Swarm

This is the method used by the Bee Inspection Service to deal with the less serious cases of EFB. If EFB is found in an apiary it is usually recommended that all the hives should be subject to a shook swarm. Beekeepers in areas where EFB is a persistent problem will often shook swarm their colonies annually or perhaps every second year as a preventative (prophylactic) measure.

The essential difference between a Bailey comb change and a shook swarm is that in the latter all the brood is destroyed and the colony starts again from scratch. If the hive has more than one brood box they both (all) have to treated in the same way at the same time. A shook swarm is best done fairly early in the season before there are any supers on the hive but, if there are supers in place, the bees can be shaken and brushed into the brood box(s)) and the boxes of comb stored securely for replacement in about two weeks' time. The most useful account of shook swarming is to be found on BeeBase -www.nationalbeeunit.com/downloadDocument.cfm?id=10.

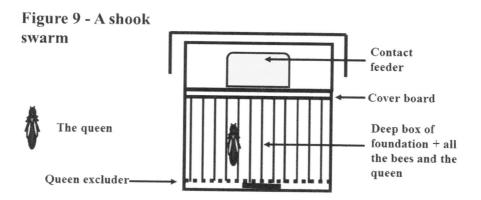

Figure 9 - A shook swarm

Comb Management

The following is a step-by-step guide to the manipulation required to do a shook swarm:-

1. Prepare a clean brood box filled with new or sterilized frames of foundation, a clean floor, a cover-board and queen excluder.

2. Set aside the colony that is to be shook swarmed and place the new floor, queen excluder and the box of foundation on the stand in that order (see **Figure 9**). The queen excluder is temporarily placed on the floor to prevent absconding

3. Remove 3-4 frames of foundation from the middle of the brood box. Remove the frames from the colony one at a time and shake and/or brush the bees off them into the gap as gently as possible. If you want to be ultra-cautious, find the queen first and put her in a safe place. A plunger type queen marking cage is ideal for this purpose.

4. Shake or brush all the bees off the old equipment (box, floor and cover-board) into the box of foundation. If you have not seen the queen be particularly careful during this operation as she may not have been on a frame. **NB** – if you have not seen the queen safely over into the box of foundation, in a few days' time, it is advisable to check that all is well and she has started to lay because, with a shook swarm, there will no brood from which the colony can raise an emergency queen.

5. Gently replace the frames of foundation to complete the new box.

6. Add the cover-board, an eke (an empty shallow box) to hold a contact feeder, followed by the cover board and finally the roof.

7. It is usually advised that you delay feeding for 48 hours. This forces the bees to use the content of their honey stomachs which may carry disease organisms. Use a strong syrup (2 parts of sugar to 1 of water) in a fairly large (4L) contact feeder because they will require at least this amount to complete their task.

8. After about a week, when brood should be present, remove the queen excluder from the floor.

9. Because all the bees have been crammed into a small volume and been generously fed, it is likely that all or most of the combs will have been drawn after a week and it is a good time to check that frames near the edge of box have been fully drawn. If not these can now be moved into positions adjacent to the current brood nest where the bees will be forced to draw them. Check that any frames you move out do not contain young larvae or eggs!

10. Keep feeding until all combs have been drawn after which the normal seasonal management can be resumed.

Shook swarming sounds like a rather harsh treatment but most colonies make a remarkable recovery. The brand new set of combs stimulates the queen to lay really hard and if it has been done early in the season (which the recommended time for shook swarming) the colony will usually produce a decent crop of honey.

Comb Management

www.ingramcontent.com/pod-product-compliance
Lightning Source LLC
Chambersburg PA
CBHW080237070525
26252CB00080B/614